WIND POWER

IAN GRAHAM

ENERGY FOREVER?

Wind Power

OTHER TITLES IN THE SERIES

Solar Power · Water Power · Fossil Fuels
Nuclear Power · Geothermal and Bio-energy

Produced for Wayland Publishers Ltd by
Lionheart Books, 10 Chelmsford Square, London NW10 3AR.

Project editor: Lionel Bender
Designer: Ben White
Text editor: Michael March
Picture research: Madeleine Samuel
Electronic make-up: Mike Pilley, Radius/Pelican Graphics
Illustrated by Rudi Vizi

First published in Great Britain in 1998 by Wayland (Publishers) Ltd
Reprinted in 2001 by Hodder Wayland, an imprint of
Hodder Children's Books

British Library Cataloguing in Publication Data
Graham, Ian, 1953-
Wind power.
1. Wind power - Juvenile literature
I. Title
333.9'2

ISBN 0 7502 3353 2

Printed and bound by G.Canale & C.S.p.A., Turin

All Hodder Wayland books encourage children to read and help them improve their literacy.

✓ The contents page, page numbers, headings and index help to locate a particular piece of information.

✓ The glossary reinforces alphabetic knowledge and extends vocabulary.

✓ The books to read section suggests other books dealing with the same subject.

Picture Acknowledgements
Cover: Getty Images. James Davis Travel Photography: 1, 11, 16 bottom. US Department of Energy: pages 13, 21, 24-25. Forlaget Flachs/Ole Steen Hansen: pages 7 top, 7 bottom, 27, 28, 30-31, 31. Forlaget Flachs: 29 (Vestas-Danish Wind Technology), 30, 33 top (Allan Flachs), Eye Ubiquitous: 8 (Andrew Cudberts), 10, 15 top (Julia Waterlow), 12 (NASA), 22 (Kevin Wilton), 24 (Keith Mullineaux), 38 (Mike Powles). Ecoscene: 4-5 (Sally Morgan), 23 (Jones), 34 (Erik Shaffer), 41 (Chinch Gryniewicz. Stockmarket/ Zefa: 16 top (Kurt Goebel). Mary Evans Picture Library: 15 inset, 17. AEA Technology: 5, 18-19, 21 inset. Samfoto: 18 (Tommy Olofsson), 34-35 (Pal Hermansen), 39 (Mira/Hatte Stiwenius), 45 (Helfe Sunde). Wayland Photo Library: 26, 36 top, 36 bottom, 37, 40 (Japan Ship Centre). OxfamUK: 33 main (Jeremy Hartley). WindStar Cruises: 42 left (Tahiti Tourism Promotions Board), 42 right, 43 (Gary Nolton).

Contents

What is wind power?

Introduction

Wind power is the use of the wind to do useful work, such as making electricity. Wind power is kinder to the environment than many other sources of energy, such as fossil fuels or nuclear power, because wind power causes little pollution. Wind is a free source of energy that can easily be put to work using windmills or their modern equivalent - wind turbines. A wind turbine looks like a two- or three-bladed propeller on top of a tall tower. There are about 200,000 wind turbines in use worldwide, but only some 50,000 of these produce electricity. The rest are used for pumping water.

Although the wind itself is free, wind power at present is more costly than other power sources. But costs are falling fast. In North America wind-generated electricity costs less than a quarter of what it did 10 years ago. Nearly half of all the world's wind-generated power is made in the United States. The next biggest producers of wind power are Denmark and Germany.

Left: Wind turbines stretch into the distance in the desert near Palm Springs, California, USA. The electricity they produce is carried away by transmission lines slung between pylons (steel towers) in the background.

FACTFILE

Wind turbines are also known as wind energy converters because they change, or convert, energy from one form – the movement energy of air – into another form, the spinning motion of a shaft, which can do work. In a car, a spinning shaft, or axle, turns the wheels. In a generator, a spinning shaft turns metal wires between magnets, and this produces electricity. Wind turbines that make electricity are also called aerogenerators. The power of a wind turbine is the rate, or speed, at which it can convert energy and do work. This is measured in kilowatts (thousands of watts).

Above: In many countries, traditional low-technology wind machines are both more affordable and more practical than high-technology wind turbines. This Cretan windmill is easy to repair using local skills and materials.

The Wind

Where does the wind come from?

Winds blow around the Earth because of the Sun. As the Earth spins, the Sun heats different parts of our home planet and it heats them unevenly. Land under a cloudless sky heats up the fastest. Clouds slow down the heating effect by reflecting some of the Sun's energy back into space. The land warms up faster than the sea because water is constantly moving and carrying the heat away. The hot surface of the Earth heats the air above it. Hot air rises. As it rises, it draws surrounding air in at ground level to replace it. These movements of air caused by the Sun are the winds.

Factfile

Clouds appear in the sky when warm, moist air cools and the invisible moisture in it condenses out into water droplets. Air cools when it rises higher in the atmosphere, where the temperatures are lower, or when it meets a stream of colder air blowing in from another area.

Wind speeds

A breeze can be pleasantly cooling on a hot summer's day, but winds can also blow incredibly fast. The strongest winds can be dangerous and destructive. Wind speed and wind direction vary at different heights above the ground. At ground level, the wind is slowed down by contact with the Earth's surface. It blows faster higher up. About 10-15 kilometres above the Earth, strong currents of air called jet-streams blow with an average speed of 140 kilometres per hour (km/h), although they can reach 450 km/h.

The diagram shows how air at ground level heats up and rises. Cool air rushes in to take it place. This produces winds.

On a sunny day at the coast, air warmed by the hot land rises and sucks in cooler air from the sea. Moisture in the sea air condenses and forms clouds as it rises. At night, the land cools down faster than the sea. The wind direction reverses as warm rising sea air sucks in the cooler air from the land.

Right: A windsock, a simple tube of fabric hanging from a pole, is a common sight at hundreds of small airfields around the world. The height and angle of the sock gives pilots a good idea of the wind's strength and direction.

How does wind affect the Earth?

Wind affects the Earth in a number of different ways. It makes waves by whipping up the sea's surface. It pushes desert sand up into dunes and moves the dunes along, grain by grain. It creates blinding sand-storms by lifting millions of tonnes of sand into the air over deserts. It spreads the Sun's heat through the atmosphere so that the Earth is warmed more evenly. It can carry clouds of air pollution hundreds of kilometres away from where the pollution was produced. It can create devastating storms.

How do plants and animals use the wind?

The smallest plant seeds and fruits are so light that they can be blown by the wind. This is an important way of spreading plants widely to different places.

FACTFILE

If the Earth had no atmosphere, its surface temperature would be similar to the Moon's. The sunlit side of the Moon is hot enough to boil water, but when the Sun sets, the temperature plummets to -173°C — which is 173 degrees below zero! Heat spreading through the Earth's atmosphere evens out these temperature extremes.

The Polish ship *Wodnica* ploughs through heavy seas near the German coast. Sailors know how the wind can transform the sea from a flat calm 'pond' into a cauldron of frothing waves.

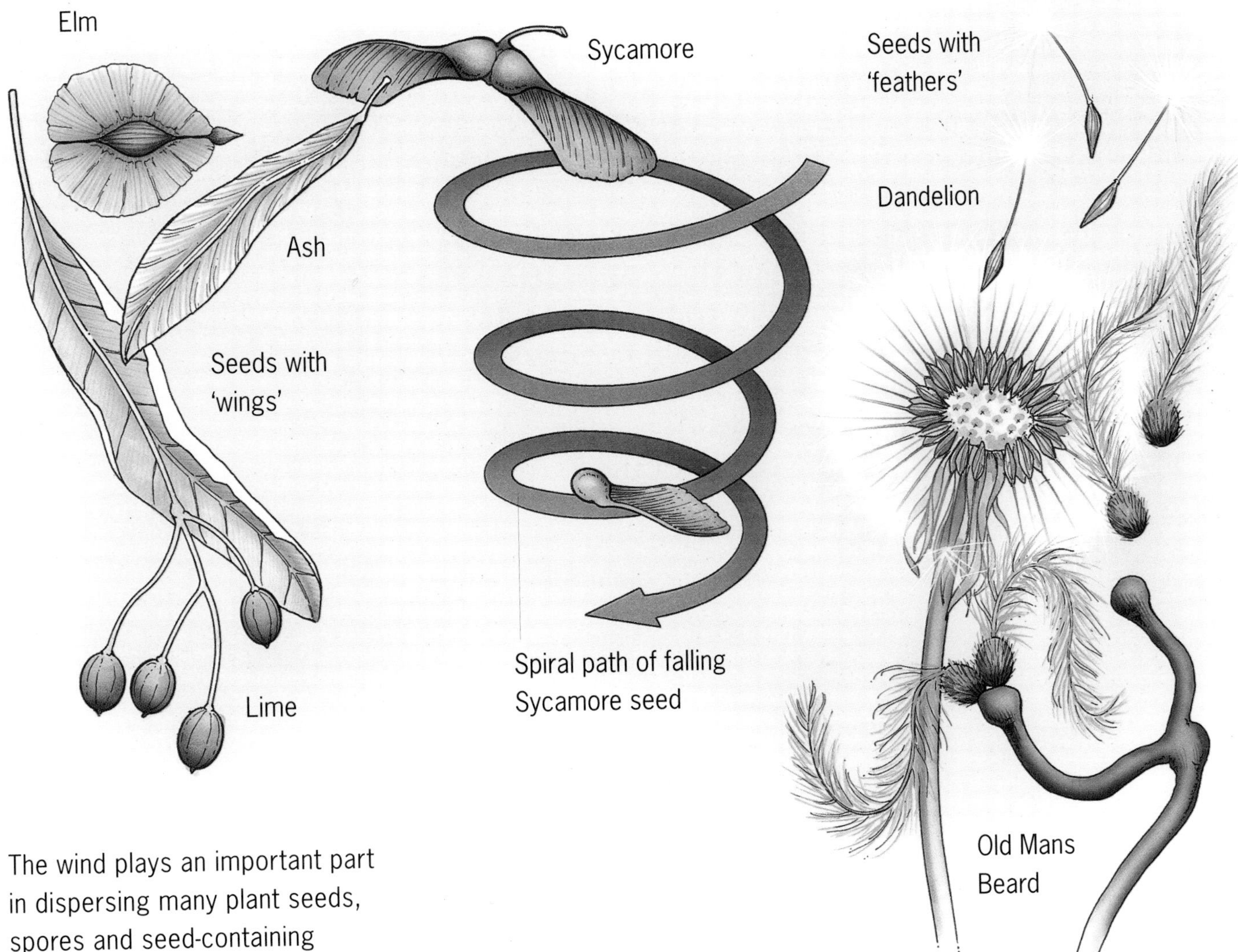

The wind plays an important part in dispersing many plant seeds, spores and seed-containing fruits. Tiny dust-like seeds and fungal spores are blown the farthest. Hairy and feathery seeds or fruits are also carried great distances. Some tree fruits have wings to slow down their fall and give the wind more time to blow them as far as 2 km from their parent tree.

By not growing next to their 'parents', new plants do not compete with them for nutrients and light. The wind carries smells long distances, too. Animals that can detect smells – they range from moths and beetles to wolves and deer – use them to help avoid predators or find food. Smells carried on the wind also enable some creatures to find their mates. A few small animals such as aphids rely on wind currents to carry them from one living area to another.

Where are winds the strongest?

Winds vary greatly in strength from place to place and from time to time according to the weather and the seasons. However, in some places average wind speeds and wind directions are very predictable.

Hundreds of years ago, sailors learned where they could find reliable winds to blow their ships across the oceans. Trade between the major continents of the world depended on these winds, which became known as trade winds. They blow from the middle of the north and south hemispheres towards the Equator, the imaginary line round the centre of the Earth.

FACTFILE

Hot air rises at the Equator, before cooling and sinking over the tropics. Some air is drawn back to the Equator, forming trade winds. The rest spreads towards the poles, forming a belt of winds called the westerlies. High pressure over the poles causes the air to move outwards, where it meets the westerlies, creating a belt of unstable weather. The Earth's rotation deflects winds to the right in the Northern Hemisphere and to the left in the Southern Hemisphere.

Kite-fliers depend on strong winds to carry their kites aloft. This kite-festival in Dieppe, on the northern coast of France, benefits from winds racing through the narrow Channel, the sea passage between France and England.

Today, we use the winds in different ways, but we still need to know where to find the most reliable winds. Wind speeds of at least 15-20km/h are needed to operate wind-powered generators and pumps. The best places to build these are mountain passes, ridges, coasts and the shores of large lakes.

A tea clipper's narrow hull and huge sail area were designed for speed. Tea clippers raced each other to deliver the first and most valuable tea cargoes from China to Europe in the nineteenth century.

The Beaufort scale

In the nineteenth century, a Bristol admiral, Sir Francis Beaufort, devised a scale for measuring wind speeds. On Beaufort's scale of 0 to 12, a breeze is in the range 2 to 6, and a gale with winds speeds of over 85km/h measures 7 or 8. Beaufort estimated wind speeds by noticing how winds affected things he could see around him. A gale-force wind sets whole trees in motion, and a hurricane whips up the sea so that the surface foams and is white with driving spray.

FACTFILE

Destructive rotating tropical storms have different names according to where they occur. They are called hurricanes in the Atlantic Ocean, cyclones in the Indian Ocean and typhoons in the Pacific Ocean.

Tornadoes and hurricanes

Tornadoes, which form over land, and ocean-borne storms produce the most extreme weather conditions at the Earth's surface. Vast spinning storms formed at sea can are called hurricanes. They can produce winds blowing at more than 250km/h. When they reach a coast they can do immense damage before friction with the land slows them down. Tornadoes spin even faster, reaching wind speeds of up to 500km/h. They can suck cars, boats and even buildings up into the air. A piece of straw blown at more than 300km/h can kill a person if it hits them.

A camera on-board a satellite orbiting the Earth captures this unmistakable view of a hurricane. It shows a giant spiral mass of clouds gathering energy over the ocean.

Above: Shattered buildings and shredded trees are strewn in the wake of *Hurricane Marilyn* in the Virgin Islands in 1995. Hurricanes are at their most dangerous as they cross the coast from sea to land.

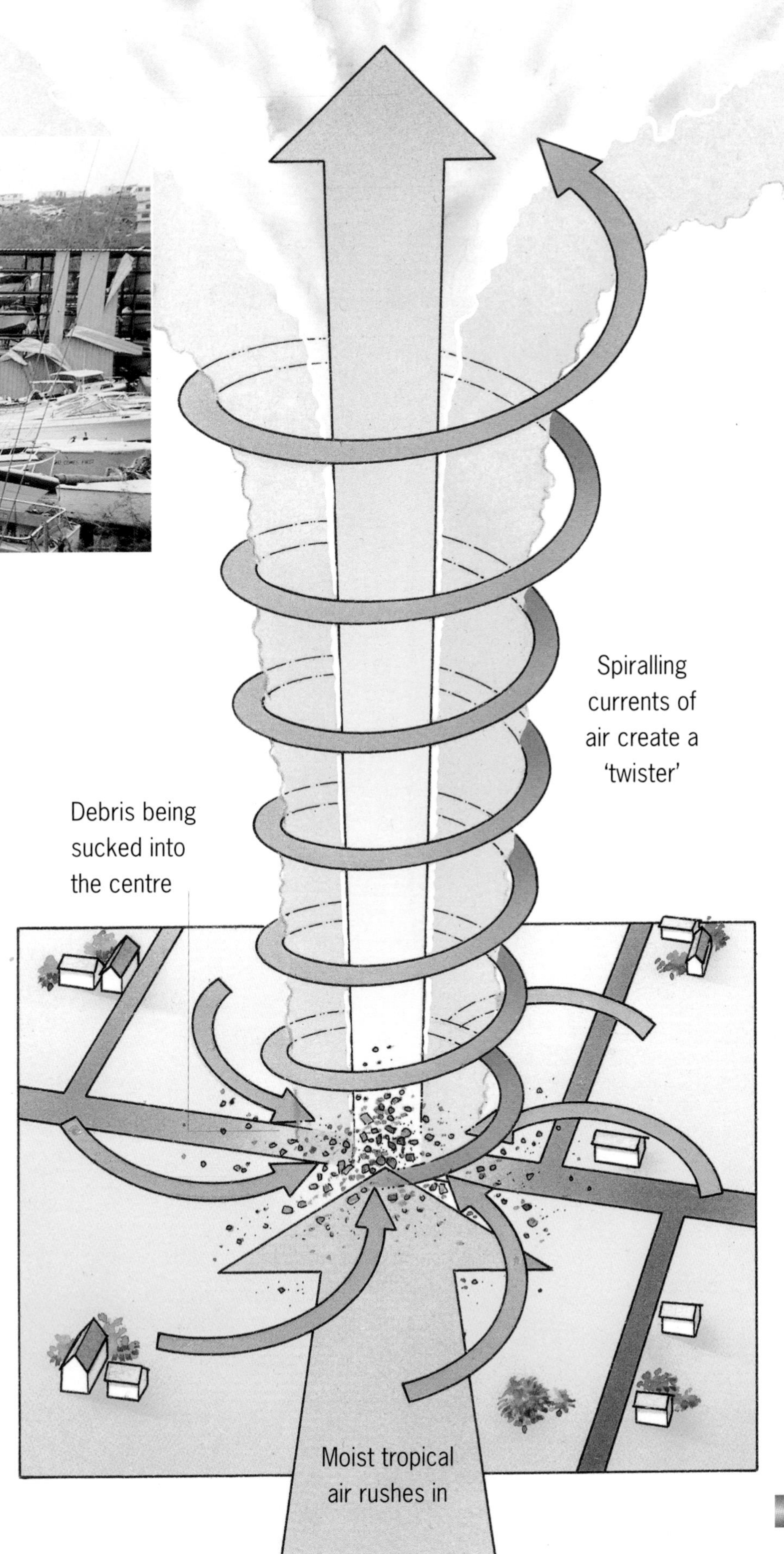

Tornadoes develop from giant flat-topped storm clouds called thunderheads. Air rises and falls and spins inside these clouds. If cool air at the base of the thunderhead sinks and carries the spinning motion down to the ground, the result is a tornado.

WIND POWER IN HISTORY

FACTFILE

The wind can have unexpected effects. In 1588, the Spanish Armada was defeated in the North Sea by the English fleet. Strong winds prevented the Spanish warships from sailing home through the Channel, so they sailed all the way round the north of the British Isles. At least 25 ships were wrecked in storms. Some of those wrecks have been found and studied. We know a lot about sixteenth-century Spanish warships because of winds that blew more than 400 years ago.

Harnessing the wind

People have made use of the wind for thousands of years. The first mechanical device that was built to use the wind as a source of power was probably the sailing boat. Sailing boats and sailing ships enabled people in different countries to trade with each other. The fleets of sailing ships used for deep-sea fishing and international trade needed harbours. These harbours attracted people looking for work and firms who could supply ships with what they needed. Large towns and cities grew up around the most important harbours. Sailing ships took explorers thousands of kilometres around the world on great voyages of discovery to new lands. They also carried conquering armies across the world.

At first, sailing vessels had simple square sails that carried them in the direction of the wind. Then Arab sailors learned how to sail into the wind using a new triangular sail called the lateen. Modern sailing yachts still use triangular lanteen sails to capture the wind in this way.

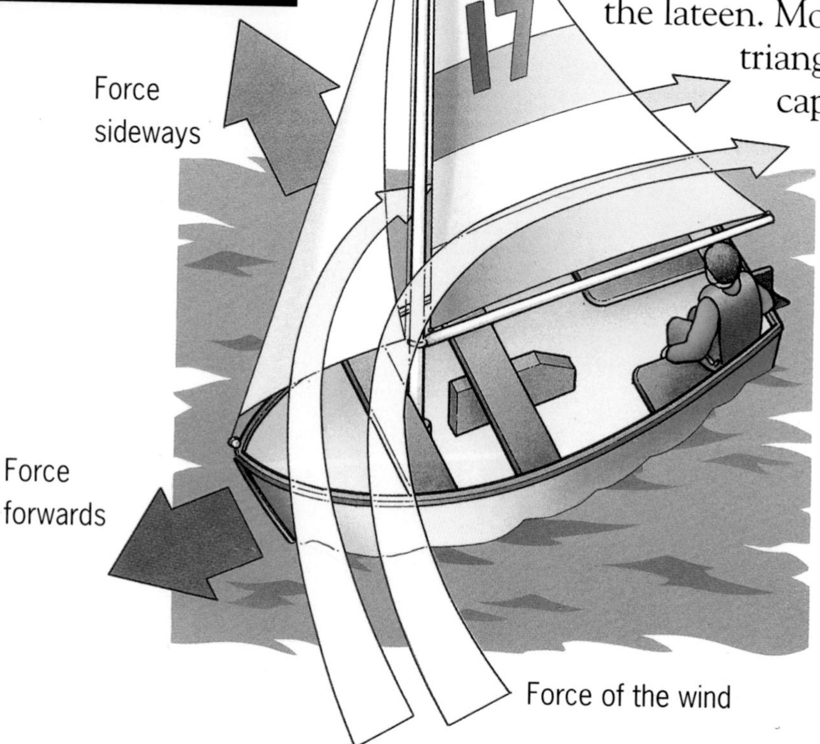

The curved shape of the lanteen sail speeds up the air blowing across it, lowering the air pressure in front of the sail. It is this drop in pressure that sucks the boat forwards.

A gentle breeze carries feluccas along the Nile at Aswan, Egypt. The felucca was developed in Spain and spread through the Mediterranean Sea to the Arab sailors of the Nile.

Left: In the twelfth century, Arab sailors invented a new type of sail for the dhow, a type of sailing boat. The lateen sail enabled the dhow to sail in almost any direction, whatever the wind direction.

Generating power from the wind

The first machine designed to use wind power to do work on land was the windmill. Windmills were unknown to any of the ancient peoples. The windmill was invented in about the seventh century in the countries that are known today as Iran and Afghanistan. From there windmills spread to the Middle East and India and eventually to China in the Far East. These early windmills were used to grind grain between heavy millstones to make flour, or were used to pump water from rivers to irrigate the land.

Inside a traditional European windmill; a large toothed wheel driven by the sails turns a small wheel linked to one of two millstones. A hopper feeds grain into a hole in the middle of the top stone. The grain is crushed between the heavy stones and comes out as fine flour for bread-making.

What did the first windmills look like?

The first windmills were quite different from the classic 'Dutch' windmill that everyone pictures in their mind. The sails, up to 12 of them, hung from a vertical post, like a boat's sails hanging from the mast and yard-arms. This design may have been inspired by a ship's sails or possibly by the wind-driven cylinders called prayer wheels that were used by Buddhists across Asia.

Windmills are still a common sight in the Mediterranean. This traditional thatched windmill is on the Greek island of Mykonos. The sail is a set of arms, like spokes on a wheel, to which triangles of cloth are attached.

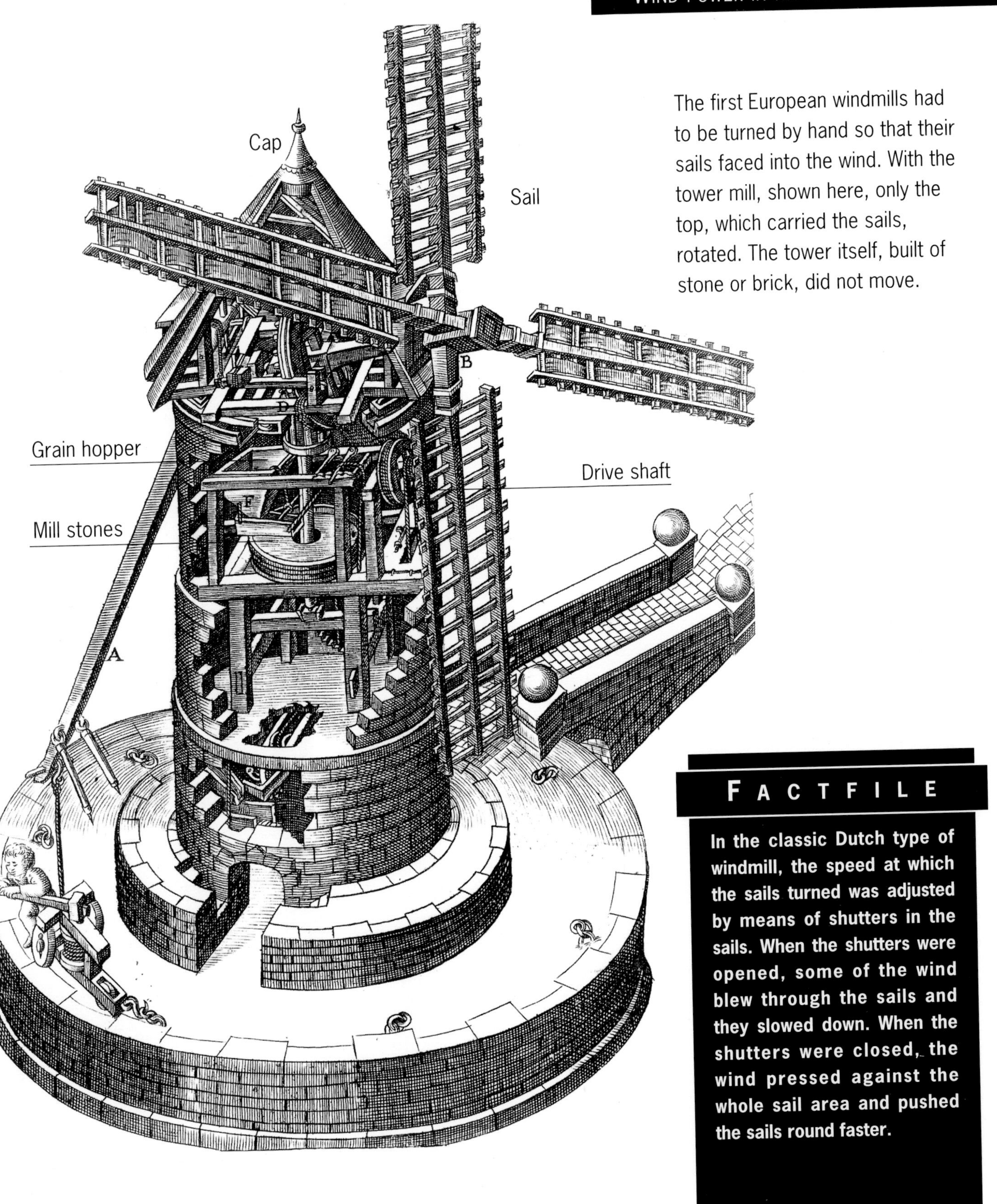

The first European windmills had to be turned by hand so that their sails faced into the wind. With the tower mill, shown here, only the top, which carried the sails, rotated. The tower itself, built of stone or brick, did not move.

FACTFILE

In the classic Dutch type of windmill, the speed at which the sails turned was adjusted by means of shutters in the sails. When the shutters were opened, some of the wind blew through the sails and they slowed down. When the shutters were closed, the wind pressed against the whole sail area and pushed the sails round faster.

The windmill comes to Europe

Windmills were not known in Europe until the twelfth century, 500 years after they appeared in the Middle East. Soldiers returning from the Crusades brought stories of windmills home with them. Europeans took up the idea of using the wind to generate power and invented a new type of mill, the post mill. The whole building could be turned about a central post so that the sails faced into the wind. Later, the simpler tower mill, which did not revolve, was developed. Only the top of the structure, where the sails were mounted, turned into the wind.

Wind power compared to water power

Windmills had two main advantages of over water mills. They did not have to be placed near running water and they could keep going when water froze in winter. Moreover, rivers where water mills were built were usually controlled by powerful landowners who could choose which people were allowed to build mills and make flour. But windmills could be built almost anywhere, so anyone could grind grain. The windmill thus began to free ordinary people from landowners.

The earliest European windmills were post mills. A post mill made for hard work because it had to be turned by hand to keep the sails facing into the wind. Very few post mills have survived to the present day. This one is in Öland in Sweden.

FACTFILE

Windmills became a very common feature of the European countryside. By the twelfth century, the Dutch were using water-pumping windmills to help them reclaim land from the North Sea. A century later, some towns in France were surrounded by as many as 120 windmills. In the Netherlands in the eighteenth century there were more than 700 windmills along the River Zaan.

The fantail, the small wheel at the top of this tower mill, was invented in the eighteenth century to turn the sails into the wind automatically. The wind turns the fantail, which drives the top of the mill round. When the sails face into the wind, the fantail is edge-on to the wind and stops turning.

WIND POWER TECHNOLOGY

How many types of wind turbine?

There are basically two types of wind turbine – with horizontal axis or with vertical axis. The main shaft of a horizontal-axis wind turbine is parallel to the ground. The main shaft of a vertical-axis wind turbine stands upright, at right angles to the ground, like the first Persian windmills.

Older wind turbines, especially those used on farms and ranches in the United States in the nineteenth century, had many blades. Modern turbines usually have two or three blades. The blades are flat, curved or wedge-shaped so that the maximum surface area can be turned into the direction of the wind.

Problems of weight

The main disadvantage of the horizontal-axis wind turbine is that the heavy gearbox and electricity generator are mounted at the top of the tower, which has to be strong enough to hold them up. Also, the equipment is difficult to reach for working on it for maintenance and repair. Vertical-axis turbines have all the heavy equipment on the ground, so the tower can be much slimmer and lighter.

Using the wind from different directions

Horizontal-axis turbines have to be turned into the wind to work at their best, so they need heavy and complex control gear. Vertical-axis wind turbines do not have to be turned into the wind. They work equally well whatever the wind direction.

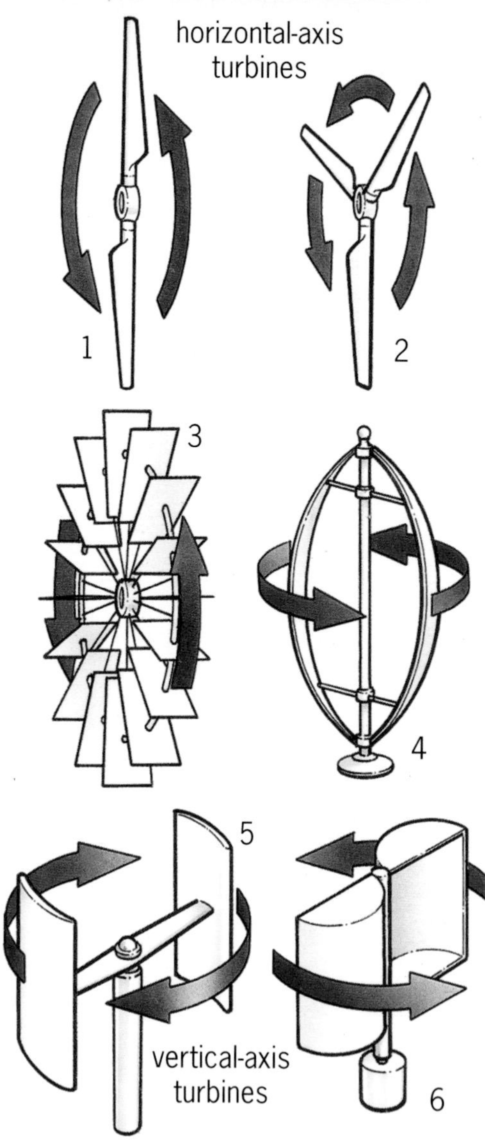

On horizontal-axis wind turbines (1, 2, 3) the blades rotate in the vertical plane. This makes it harder for them to work unless they are facing into the wind. Vertical-axis wind turbines (4, 5, 6) which rotate in the horizontal plane, can catch the wind much more easily.

FACTFILE

The egg-beater type of wind turbine was invented in the 1930s in France by George Darrieus, but Darrieus did not develop his imaginative design. Canadian engineers reinvented the design and developed it to produce a practical power-generating wind turbine in the late 1960s.

The modern vertical-axis wind turbine, such as this cross-arm rotor, looks quite unlike the traditional windmill. Instead of broad sails facing the wind, it has tall thin wing-like aerofoils.

One type of vertical-axis wind turbine, the Darrieus rotor, is also known as the egg-beater because it resembles a common kitchen egg whisk.

Using the basics

In many parts of the world, high-technology wind turbines are too expensive for most people to construct and use. However, wind power can be generated with very simple, low-technology equipment and basic, local materials. The ability to make wind turbines cheaply and easily from everyday materials – the practice of 'appropriate technology' – is vital to their use becoming more widespread in developing countries.

Below: New types of wind turbines are still being developed. This curious-looking structure is an experimental wind turbine being tested by National Power, Britain's electricity transmission company.

The bicycle wheel wind turbine at the Centre for Alternative Technology near Machynlleth in Wales generates 1 kilowatt-hour (kWh) of electricity every month, which is used for charging batteries.

Do-it-yourself wind turbines

A wind turbine can be made with something as basic as an oil drum or a bicycle wheel. The generator that makes the electricity when it is turned by a wind turbine need not be specially designed and made for the job either. Every car has an alternator, or generator, to charge the battery. The car's engine drives the alternator, which generates electricity to keep the battery fully charged. Alternators are readily available from old cars that are no longer roadworthy.

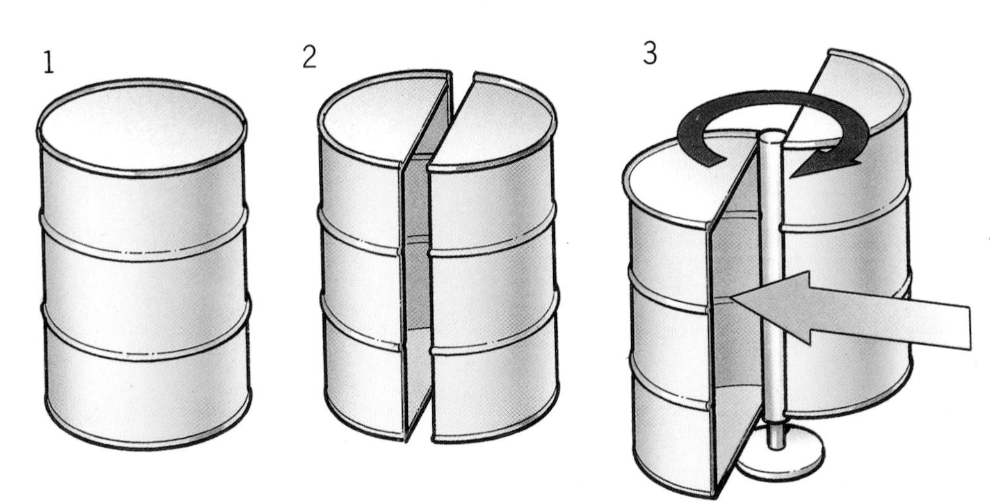

One type of vertical-axis wind turbine, the Savonius rotor, can be made from an oil drum (1). The drum is cut in two (2) and the two halves are welded to a steel pipe.

The pipe is then slipped over a pole so that it can rotate. Whatever the wind direction, wind always fills the open half of the drum and pushes it round (3).

FACTFILE

The principle of the Savonius rotor is used by a type of spinning advertising sign. Instead of being flat, the sign is bent into the shape of a letter 'S' standing on end - like the Savonius rotor. As the sign spins in the wind, the advertisements on both sides appear one after another.

How safe are wind turbines?

Engineers are always looking for ways of making wind turbines more efficient, so that they can squeeze more power out of lighter winds. Of course, as the wind speed increases, so does the turbine's. Strong winds could make it spin so fast that the vibrations and powerful forces produced could tear the blades off the turbine altogether. It is important, therefore, that wind turbines have a built-in safety feature that automatically limits their maximum speed.

How are wind turbines stopped?

Large commercial turbines have electronic control systems that measure the wind speed and direction. The angle of the blades and the direction of the whole turbine are constantly adjusted to suit the wind conditions. When the wind blows too strongly, the blades are turned out of the wind so that they cannot spin too quickly. Large wind turbines are also fitted with brakes to slow them down. Smaller turbines may have weights attached to the blades. As the blades spin faster, centrifugal force – a force acting outwards from the centre – pushes the weights outwards. This twists the blades so that they cannot catch the wind so easily, slowing them down. Turning blades out of the wind is called feathering.

Engineers work on an experimental wind turbine located at the foot of the Colorado Rocky Mountains. This is an area that is known for variable winds and winter storms – an ideal testing ground for wind turbines.

A wind turbine's blades must be securely fixed to the central hub, not only to support their weight but also to withstand the powerful forces produced by their spinning motion. When the blades spin, forces acting outwards try to pull the blades off.

FACTFILE

The world's first really big wind turbine was built in 1941 by the American engineer Palmer Puttnam on top of a mountain called Grandpa's Knob in Vermont, USA. The two-bladed rotor was 53m from tip to tip and sat on top of a 33 m-high tower. It generated 1.25 megawatts (MW) of electricity. It failed when one of its massive blades flew off in March 1945. Many other wind turbines built at that time suffered the same fate, as engineers learned to cope with the forces produced by spinning turbines.

Effects of wind power on the environment

Wind turbines are kinder to the environment than other power schemes. For example, they don't produce any harmful waste. However, people have voiced two main objections to them – noise and appearance.

Swishing blades and machinery noises

People who live close to wind turbines have complained about the nosily swishing of their blades and the grinding noise made by their gearboxes and generators. The latest wind turbines are much quieter than older models. Blade swishing cannot be heard more than about 200 metres from a turbine and engineers say that mechanical noise has been almost eliminated from the latest turbines.

Animals seem happy to share the countryside with wind turbines. These sheep are grazing contentedly under a line of wind turbines, untroubled by blades spinning above their heads.

Sharing the land with wind turbines

Some people feel that wind farms are ugly and destroy beautiful views of the countryside. In some places, the towers and turbines can also cause interference with television reception. Wind turbines certainly occupy large areas of the landscape, but the towers themselves take up very little space at ground level. The land between the towers can be used for other purposes – usually farming. Cows or sheep can share the land and graze right up to the bases of the towers. They seem to be unaware of the blades whirling round above them.

FACTFILE

Birds have been killed in and around wind farms. The deaths of rarer species such as eagles are particularly worrying. However, most bird deaths seem to be caused by collisions with the power lines, which are difficult to see, not with the wind turbines themselves. Studies carried out in the Netherlands, Denmark and the United States found that the numbers of birds killed by wind farms are tiny compared to the numbers killed by traffic.

The Danish company Vestas has set up more than 600 wind turbines in India. Local people are trained in how to build, use, maintain and repair the turbines. The open steel towers that support the turbines are much easier to make and erect than concrete towers.

USING WIND POWER

Making electricity from the wind

A wind turbine on its own does no useful work. It just spins in the wind. To produce electricity, it has to be coupled to a generator. The generator changes energy from the wind into electrical energy. The spinning part of the generator, turned by the wind turbine, moves coils of wire past magnets to make electricity flow in the wire. Electricity and magnetism are very closely linked. When electricity flows along a wire, it produces magnetism too. And the opposite is also true – when a magnet moves near a wire (or a wire moves near a magnet), it causes electricity to flow along the wire.

Below: An engineer in Denmark checks an important part of a wind turbine – the ring of bolts and fittings that connect one of the blades to the massive central hub – before the turbine leaves for India, where it will be assembled for use.

Getting the speed right: using a gearbox

In the case of a small wind turbine supplying electricity to one home, the turbine can be connected directly to the generator. In the case of larger turbines used by electricity supply companies, the blades turn too slowly to drive the generator fast enough. The answer is to fit a gearbox – a mechanism, consisting of toothed wheels, that acts between the turbine and the generator to change the speed. Hence the shaft coming out of the gearbox spins faster than the turbine. This shaft drives the generator.

FACTFILE

The world's largest wind turbine is a Boeing Mod-5B in Oahu, Hawaii. Its rotor blades are 97.5m across. It can generate up to 3.2 megawatts in a wind blowing at 50km/h.

The generator cabin of the largest wind turbines is big enough to climb into and walk around inside. It houses the generator, gearbox and control equipment.

Above: A crane prepares to hoist a wind turbine's nacelle (generator housing) to the top of its tower during the construction of a wind farm in Germany. An already completed wind turbine stands in the background.

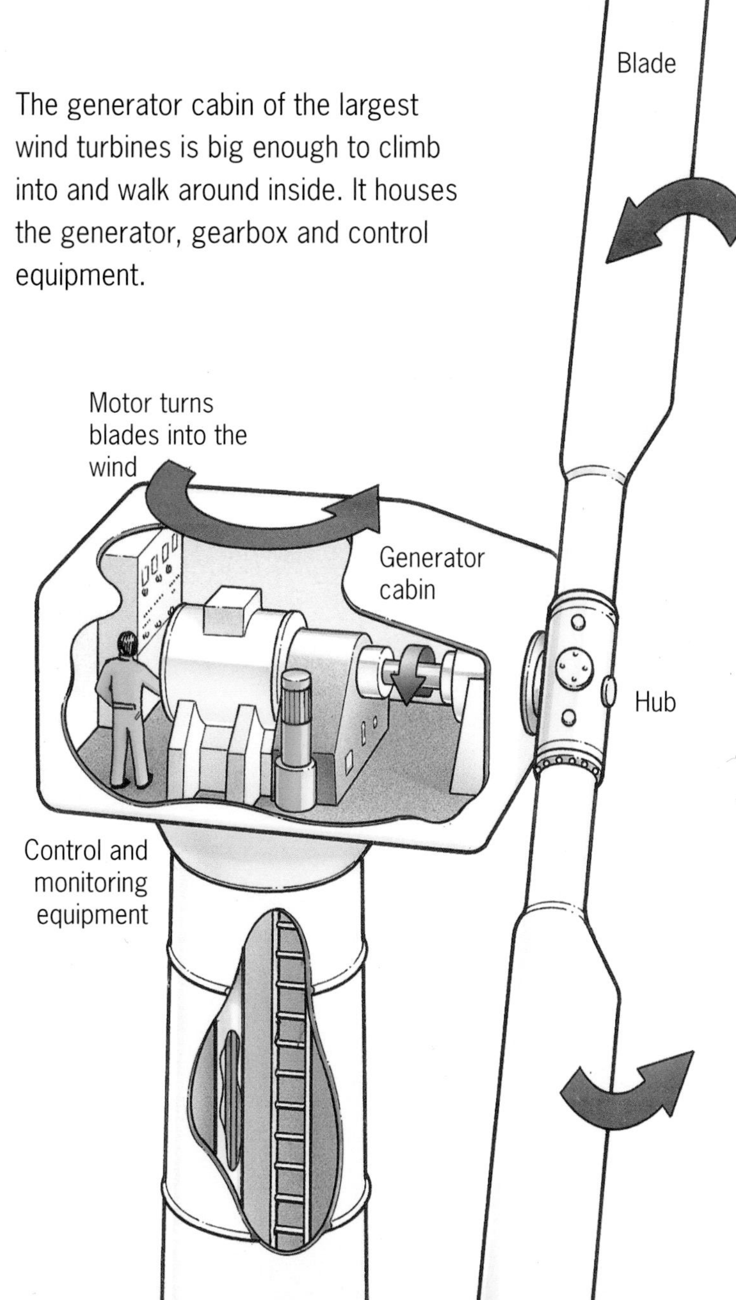

Wind power for local communities

More wind turbines are used in Denmark than anywhere on Earth apart from California, USA, and Germany. Denmark's 3,700 wind turbines generate 1,000 million kilowatt-hours of electricity per year, which is about one-thirtieth of the country's electricity consumption.

Unlike California, where wind turbines are usually concentrated in large numbers on wind farms, three-quarters of Danish wind turbines are installed singly or in small numbers to supply local communities. The turbines are often owned by the people who live close to them. Nearly 50,000 people out of Denmark's population of five million own a wind turbine or have a share in one.

Exporting wind-power technology

Denmark is also the world's leading exporter of wind turbines. About half of all international trade in wind turbines is conducted by Danish companies. Danish wind turbines operate in countries as diverse and far-flung as Switzerland, China, Israel, New Zealand and Mexico.

Transmission lines carry electricity away from the Studstrup coal-fired power station in Denmark. Even here, where fossil fuel power is plentiful, wind turbines have been erected to generate clean power.

Wind turbines line the water's edge in the harbour of the Danish fishing village of Bønnerup. Coasts are good places for wind turbines to catch winds which often blow inland off the sea.

Right: A de Havilland Tiger Moth biplane, an example of technology from a bygone era, gives its pilot a bird's- eye view of a Danish wind farm built using the latest technology.

Wind power at home

A small wind turbine can be set up outside someone's home to generate enough electricity for their private use. In remote areas, a windmill might be the only way of producing electricity cheaply without equipment such as a petrol- or diesel-powered generator, or high-technology equipment like solar panels.

Lighting up the Wild West

From the middle of the nineteenth century, American farms and ranches often had a many-bladed windmill, called a western wheel, to pump water from a river or lake onto the land. A few of those original old windmills are still in working condition. In the 1920s and 1930s, farms and homes in remote areas of the United States with no public electricity supply used 200-3,000 watt wind generators to power their lights and perhaps a radio.

Wind power for the home today

A typical home wind turbine today has aluminium or fibre-glass blades set on top of a tubular steel tower. As the blades turn, they drive a generator – possibly salvaged from an old car. A small home wind turbine of this kind can produce 750 watts of electrical power – enough for lights or small appliances.

FACTFILE

The plans for the first Danish eco-village were drawn up in 1981. The land for it, covering 13 hectares, was bought at Torup in 1988. The first 14 houses were built there in 1991. Electrical power is supplied by solar panels and a 450-kilowatt wind turbine.

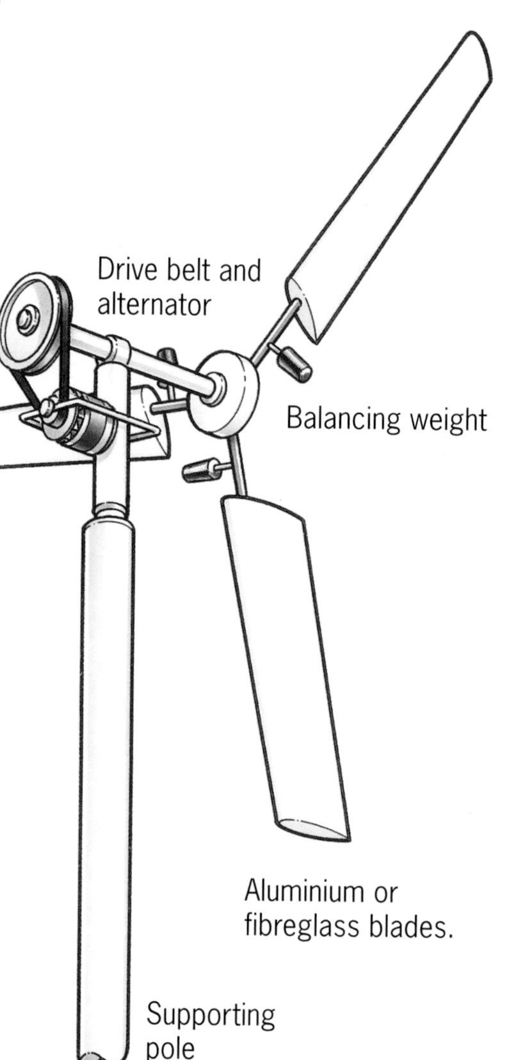

Wind turbines small enough to supply power to just one house can be bought as kits or built from scratch by do-it-yourself enthusiasts. The electricity generated by this model is carried down to ground level by cables inside the pole that supports the turbine.

Above: Eco-villages, such as Torup in Denmark, shown here, use energy-efficient houses to reduce the community's electricity needs. The little electricity that is needed is generated by environmentally friendly methods including solar and wind power.

Right: A wind turbine pumps water for people and animals in an African village. Even in this apparently dry, lifeless landscape, water may be quite close to surface. Harnessing a little power from the wind can transform life in a remote village like this.

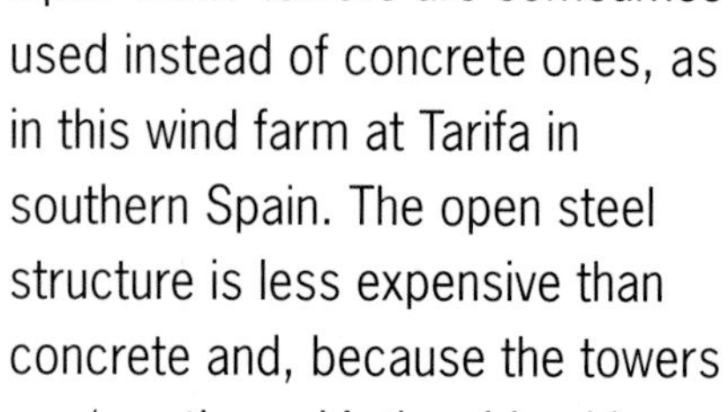

Open trellis towers are sometimes used instead of concrete ones, as in this wind farm at Tarifa in southern Spain. The open steel structure is less expensive than concrete and, because the towers are 'see-through', they blend in with the countryside better.

Wind farms

The wind turbines used by the electricity industry are built in groups called wind farms, or wind parks. Compared to the turbines used by individual homes, they are massive. Their rotor blades are about 20-30 metres across, though some are much bigger, and they stand on towers up to 50 metres high.

One of these wind turbines on its own generates about 500 kilowatts of electricity (1 kilowatt is enough for a one-bar electric fire), but when hundreds or thousands of them are operating together, they can generate hundreds of megawatts (hundreds of millions of watts) of electricity – enough to power a whole town. However, wind farms need a huge amount of space.

Getting the spacing right: how far apart?

If wind turbines in a wind farm are placed too close together, they 'shade' each other – just as overlapping sunbathers would shade each other from the sun. Their positions have to be worked out very carefully so that they can be as close together as possible to extract the most wind power from the smallest area without shading each other. Wind turbines are usually separated by between five and seven times their rotor diameter.

FACTFILE

Although wind power is a clean form of energy, building a wind farm can have a major impact on the environment. The foundations for the towers have to be sunk up to 50 m deep, and rocky soil is sometimes blasted with dynamite. Often, digging huge holes disturbs the ecology of the site, so that the plants that grow there afterwards are different from those that grew before. Also, the tracks built to reach the site can cross fragile natural habitats. In Yorkshire, England, moorland habitats supporting rare species have been threatened in this way.

This wind farm in Bornholm, Denmark, is built on a gently sloping hillside. Hills are often good sites for wind farms because of the way the prevailing wind sweeps up the slope on one side.

Ranks of wind turbines stand in the Tehachapi-Mojave wind farm in California, USA. Although the Altamont Pass contains more wind turbines, Tehachapi-Mojave produces more electricity. The generator housing of one of the turbines in the foreground is open for servicing.

Different types of wind turbine are continually being refined and tested. At this American test facility, two vertical-axis Darrieus 'egg-beater' turbines stand next to a massive three-blade horizontal-axis rotor.

Altamont Pass, California, USA

The Altamont Pass in California, USA, just one hour's drive east of San Francisco, contains the world's largest concentration of wind turbines. Six thousand of them produce almost half of all the wind power generated in California. Together, they produce one terawatt-hour (1,000 million kilowatt-hours) of electricity per year. Mountain passes are good sites for wind farms because the wind is funnelled and squeezed between the mountains, increasing its speed and force. The Altamont Pass is especially good because it lies in the hills between the San Francisco Bay area and Pacific Ocean on one side and the hot San Joaquin Valley on the other side. There, air rising from the hot valley sucks in cooler air from the sea through the pass to create winds. Standing in the pass, the wind turbines tap this energy and turn it into electricity.

Just over half of the Altamont Pass wind turbines are a type known as downwind rotors. Most wind turbines are designed to hold their rotor blades in front of the tower. Downwind rotors have their blades behind the tower.

Wind turbines are a little like icebergs in that some of their structure is out of sight below the surface. With the weight of a heavy generator to be held steadily high above the ground, the tower has to be stable and firmly rooted in a solid, deep foundation.

A traditional windmill pumps water on the Lasithi Plateau on the Greek island of Crete. If the wind direction changes, wind pressure pushing against the triangular vane swings the fabric sails round until they are pointing straight into the wind again.

The need for pumping water

Most of the world's wind turbines are used to pump water to irrigate the land. In many parts of the world, it is either too dry all year round to grow crops or there is a rainy season for a few months and the rest of the year is quite dry. Valuable crops are raised during the wet season, but for much of the year there is not enough water to grow those crops. Only grass for animal feed can be grown in the dry season. However, there is sometimes a plentiful supply of water lying under the ground that could be used to irrigate high-value crops year round, if only the farmers could reach it.

Wind-powered pumps

Wind turbines can drive pumps to bring this water up from underground and move it to where it is needed. In fact, underground water was pumped up to the surface long before wind turbines became available. People used kerosene-powered pumps, which develop their power by burning kerosene, a cheap fuel. Such pumps are cheap to buy and to run, but they are inefficient and do not last very long. Wind turbines cost more to set up but, once installed, they are more reliable and have lower running costs because they do not burn any fuel.

FACTFILE

The efficiency of wind-powered pumps for bringing up water from underground depends on the level of the 'water table'. This is the upper level of water that has collected below the surface in porous rocks. As the water table falls, the pumps must work harder. They can do so only if the rotor blades are turned faster by the wind. The water table rises when it rains. It falls during dry periods – and as water is pumped up from underground.

This many-bladed fantail turbine, pictured in Santa Cruz, Argentina, is typical of the type of wind turbine used for pumping water on American farms and ranches at the end of the nineteenth century. These 'western wheels', as they are popularly known, are still in use all over the world.

Modern wind-powered vessels

Thousands of years after the first sailing ships took to the sea, the wind is still used to propel boats and ships. Today, this is mainly for sport and leisure, but there is also some interest in using wind power for commercial shipping. Yachts, sailing dinghies, sail-boards, land yachts and ice yachts are all wind-powered craft.

A few passenger liners, oil-tankers and other large commercial ships have also been equipped with sails. They use their sails when the wind is strong enough, but they switch to engine power when the wind drops.

Using sail power with an engine as a back-up power source cuts costs and pollution by reducing the amount of fuel that has to be burned. It can also cut voyage times. An offshore gas rig fitted with two 60-metre sails cut its expected journey time from Louisiana across the Atlantic Ocean to the North Sea by five days.

The cargo vessel *Tropical Marina* raises its sails and prepares to cut its engines. By hitching a free ride until the wind drops, the ship burns less fuel, which is expensive, and helps to create a cleaner environment.

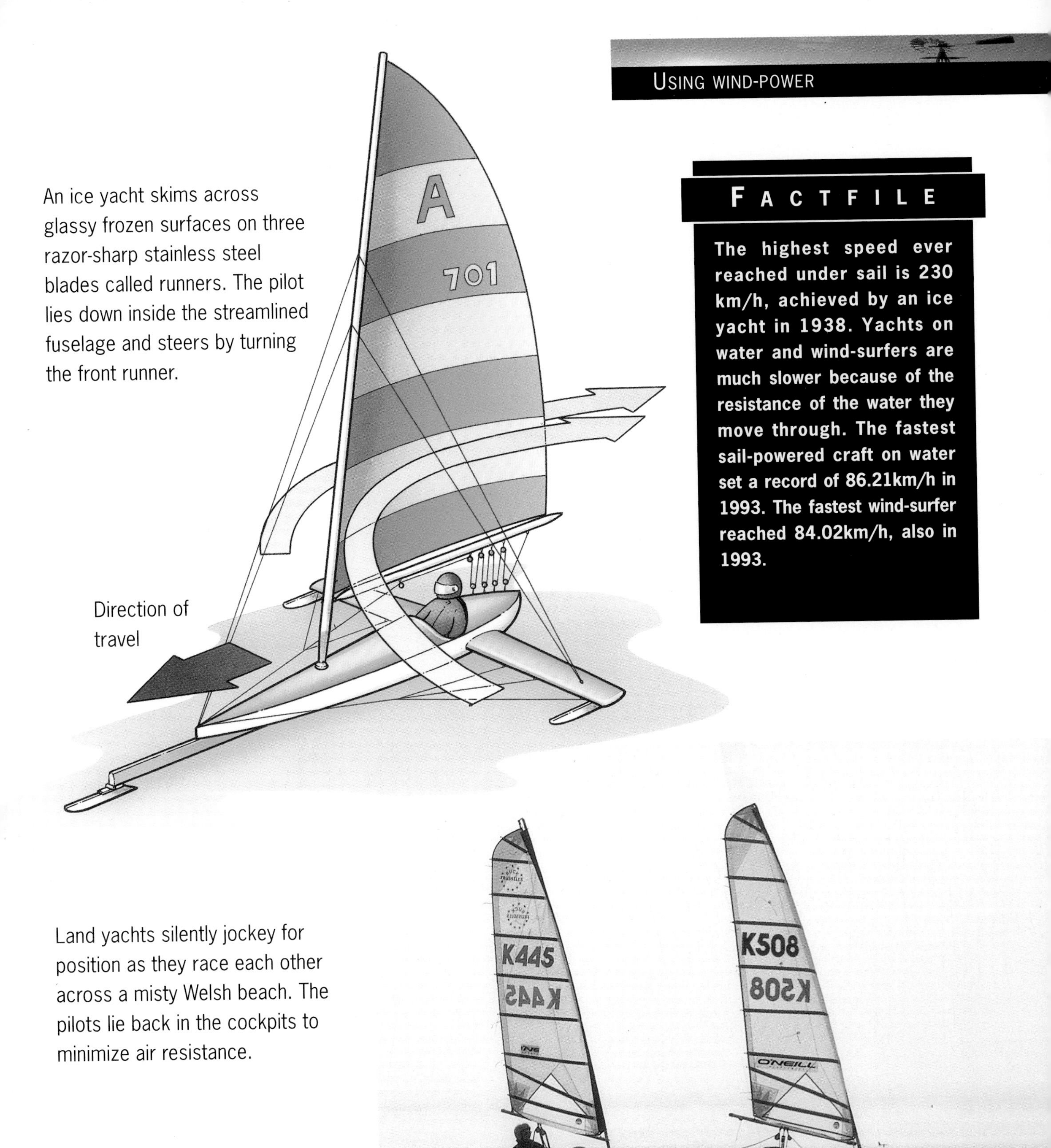

An ice yacht skims across glassy frozen surfaces on three razor-sharp stainless steel blades called runners. The pilot lies down inside the streamlined fuselage and steers by turning the front runner.

FACTFILE

The highest speed ever reached under sail is 230 km/h, achieved by an ice yacht in 1938. Yachts on water and wind-surfers are much slower because of the resistance of the water they move through. The fastest sail-powered craft on water set a record of 86.21km/h in 1993. The fastest wind-surfer reached 84.02km/h, also in 1993.

Land yachts silently jockey for position as they race each other across a misty Welsh beach. The pilots lie back in the cockpits to minimize air resistance.

A wind-powered luxury ship

Club Med 1 is one of the world's biggest sailing ships. From bow to stern, it measures 187 metres and it weighs some 14,000 tonnes. *Club Med 1* is not a hangover from a bygone golden era when sailing ships ruled the oceans. It is a brand-new luxury passenger liner designed from scratch to make use of wind power. It was launched in France and made its first, or maiden voyage in 1990.

Computerized sailing

Club Med 1 can raise seven sails covering an area of 2,700 square metres on five masts. The masts, which stand 50 metres high, are made from aluminium, one of the lightest metals, to save weight. Their hollow tubular shape gives them the strength to withstand the bending forces produced by wind pressure pushing against the sails. Computers continually monitor the sails and adjust them to get the best performance from them. Despite the weight of the huge masts and sails and their control gear, the bottom of the ship's hull is less than five metres below the water's surface, so *Club Med 1* can enter many small harbours that larger passenger liners cannot.

Below: Seen from above when its sails are down, *Club Med 1*'s masts and booms make a matchstick-like array. The luxury liner has accommodated up to 425 passengers in cabins on eight decks. Its latest refit will have more spacious cabins and suites, and the ship is to be renamed *Wind Surf*.

Left: From the deck of *Club Med 1*, passengers get a magnificent view of the sails towering above the ship's superstructure.

Dolphins break the surface within sight of the motorized sailing ship *Wind Song*. This and *Club Med 1* are two of a fleet of modern purpose-built passenger sailing ships that mainly cruises around the islands and popular holiday destinations of the Mediterranean and Caribbean Seas.

FACTFILE

Sailing ships used to need teams of sailors to climb the masts and unfurl the sails, or haul extra sails up the masts. Modern sailing ships like *Wind Song* and *Club Med 1* use computer-controlled power winches (electric motors) to operate the sails.

The Future

Wind power in the future

Wind power will continue to grow as people become more concerned about atmospheric pollution. Canada, for example, is planning to generate up to one-fifth of its electricity from the wind. One of the drawbacks of wind power in the past was its cost. But now, electricity from the best wind turbines is as cheap as electricity from a coal-fired power station. Researchers believe that wind power can be reduced to half of its current cost by using lightweight composite materials, better computerized control systems and more efficient generators.

Below: An artist's impression of large-scale use of wind power in the future. As all forms of renewable energy grow in importance in the twenty-first century, wind turbines will be seen more and more in towns, on hillsides and along coasts.

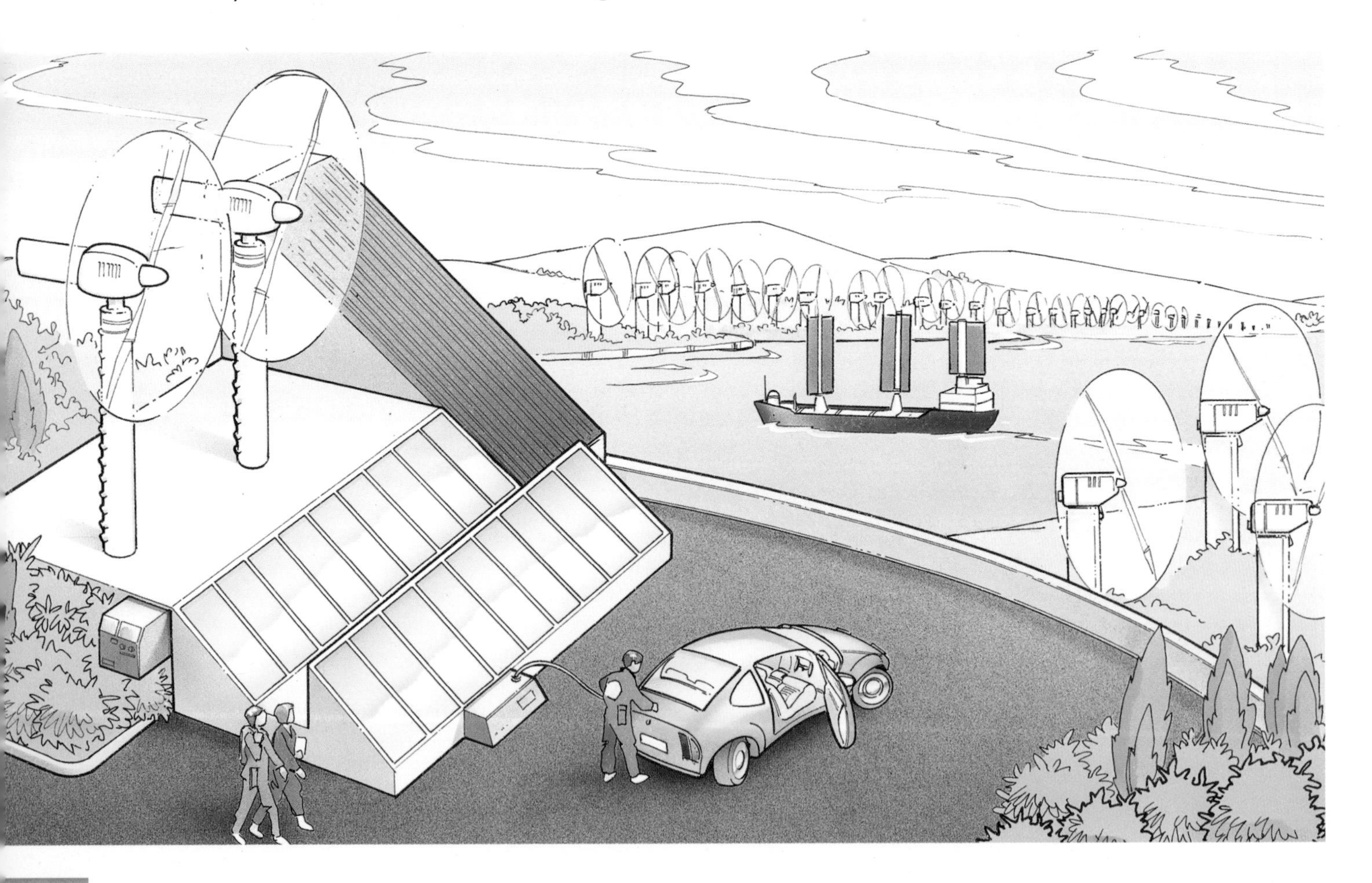

FACTFILE

Building offshore wind-power plants presents problems that land-based wind farms do not suffer from. The sea wind carries salt spray, which is very corrosive and eats into metal parts. The wind turbines at the Vindeby offshore power plant in Denmark are protected against salt corrosion in two ways. First, their towers are airtight. Second, the air pressure inside their nacelles (generator casings) is kept higher than normal, to stop moist salty air from seeping in.

Wind turbine manufacturers are designing modular, or standardized unit, structures that will make assembly quicker and easier and reduce costs. If these developments can be achieved, more and more wind turbines will spring up around the world. Wind turbines standing on towers near homes and farms, especially in remote areas, could become a common sight in the years ahead.

Harvesting the sea winds

Wind farms might also be built offshore, in the sea, where the winds are often stronger and less turbulent. In fact, the world's first offshore wind farm already exists. In 1991, 11 wind turbines were built in the sea 1.5 kilometres off the Danish coast near Vindeby on Lolland Island. Together they produce a total of 5 megawatts of electricity.

Denmark has the most ambitious wind-power programme in Europe. More wind farms like this one at Ebeltoft will be built over the next 30 years. By then, about 40 per cent of Denmark's electricity needs should be met by wind power.

GLOSSARY

Aerogenerator A wind turbine used to produce electricity.

Alternative technology Using machines that produce less pollution and so are kinder to people and the environment.

Alternator An electricity generator used by cars and other vehicles.

Blade swishing The rushing noise made by a wind turbine's blades cutting through the air.

Centrifugal force A force on a revolving object acting outward from the centre.

Condense Change from a vapour or gas into a liquid.

Cross-arm rotor A type of vertical-axis wind turbine.

Cyclone A powerful rotating storm in the Indian Ocean.

Darrieus rotor A type of vertical-axis wind turbine shaped like an egg-beater.

Eco-village A village of energy-efficient houses, designed to need as little electricity as possible.

Environment The world around us – landscape, air, plants, animals, rivers, seas.

Fantail A small wheel or vane to one side of a windmill, used to turn the windmill into the wind automatically.

Feathering Turning a wind turbine's blades out of the wind, so that the turbine does not turn too fast in strong winds.

Fossil fuel A fuel such as coal, oil or natural gas formed from the remains of microscopic plants and animals that lived millions of years ago.

Generator A machine for changing movement energy into electricity.

Hurricane A powerful rotating storm in the Atlantic Ocean.

Kilowatt (kW) One thousand watts, a measure of electrical power.

Kilowatt-hour (kWh) A unit of energy equivalent to 1,000 watts of electrical power being used for one hour.

Megawatt (MW) One million watts, a measure of electrical power.

Post mill A type of windmill in which the whole building has to be turned about a central post to face the sails into the wind.

Power station A building where energy released from a fuel is converted into electricity.

Pressure A force exerted by one thing on another in contact with it.

Rotor A set of blades attached to a central hub so that they can rotate.

Tornado A narrow tube of whirling winds blowing at great speed.

Turbine Angled blades fitted to a shaft that is free to rotate. A moving gas or liquid pressing against the blades makes the turbine rotate.

Typhoon A violent rotating storm over an ocean.

Watt (W) A unit of electrical power, a measure of how quickly electrical energy is being converted into other forms such as heat.

Windmill A building with wind-driven sails that turn a millstone to grind corn into

FURTHER INFORMATION

Books to read

Cycles in Science: Energy by Peter D. Riley (Heinemann Library, 1997)
Earthcare: Raw Materials by Miles Litvinoff (Heinemann Library, 1996)
Eyewitness Science: Energy by Jack Challoner (Dorling Kindersley and London Science Museum, 1993)
Power from the Wind by Hazel Songhurst (Wayland 1994)
Science Topics: Energy by Ann Fullick and Chris Oxlade (Heinemann Library, 1998)
Science Works: Energy by Steve Parker (Macdonald Young Books, 1995)
The Super Science Book of Energy by Jerry Wellington (Wayland, 1994)
Wind Energy by Graham Rickard (Wayland 1995)
The World's Energy Resources by Robin Kerrod (Wayland, 1994)

Power and energy consumption

Power is the measurement of how quickly energy is used. It is measured in joules per second, or watts. An electric iron might need 1,000 watts to work, but a portable radio might need only 10 watts. The energy needed to keep the radio going for one hour would run the iron for only six minutes, because the iron uses up energy ten times faster than the radio. The diagram to the right compares the power ratings of household electrical goods and of homes and power stations.

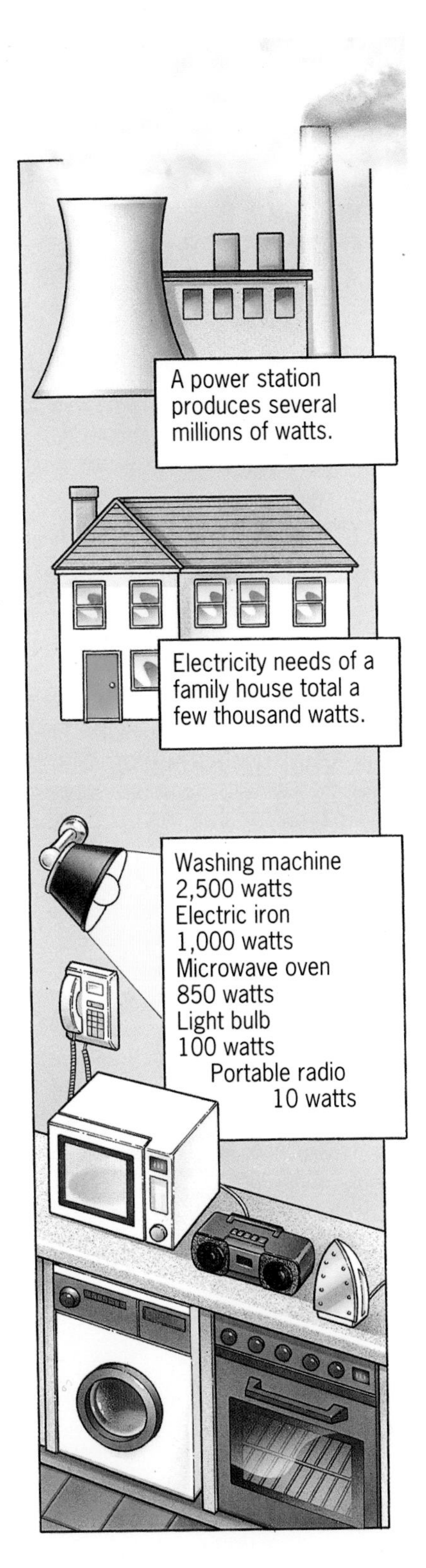

INDEX